FUN WITH SCIENCE

CHEMISTRY

STEVE PARKER

Contents

Use the symbols below to help you identify the three kinds of practical activities in this book.

EXPERIMENTS

TRICKS

THINGS TO MAKE

Illustrated by Kuo Kang Chen · Peter Bull

WARWICK PRESS

Introduction

Everything in the world is made of chemicals. Rocks, soil, houses, bridges, cars, plants, and animals—and you. Everything is made from a set of basic chemical substances. Chemistry is the scientific study of how these chemicals are joined to form the objects around us, and how we can split or combine chemicals into new substances.

The questions on these two pages are based on scientific ideas explained in this book. As you carry out the experiments, you can answer these questions, and understand more about how chemicals make up our world.

Safety and equipment
The activities in this book use only common household substances and pieces of equipment. Keep chemicals safe in containers. Wash your hands before and afterward. Never put chemicals near your face or mess about with them "to see what happens." This is neither safe, nor good science.

This book covers eight main topics:
- How chemicals make up the world
- Atoms and molecules, building blocks of chemicals
- How chemicals react together
- Colored chemicals
- Chemicals as solids, liquids, and gases
- Chemicals that are metals
- Chemicals containing carbon
- Chemical factories, waste, and pollution

A blue line (like the one around the edge of these two pages) indicates the start of a new topic.

▲ What is an acid, and how is it different from a base? (pages 16–17)

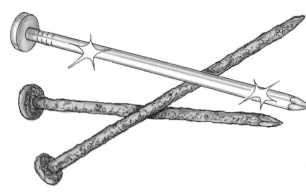

▲ What is a chemical, and what makes it change into another chemical? (pages 4–5)

▼What makes a chemical change from a solid, to a liquid, to a gas? (pages 20–23)

Pressure of skate blades on ice melts it into water.

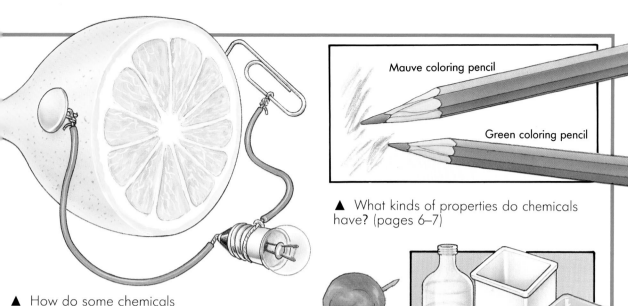

Mauve coloring pencil

Green coloring pencil

▲ What kinds of properties do chemicals have? (pages 6–7)

▲ How do some chemicals react together to make electricity? (pages 28–29)

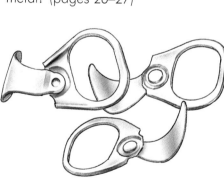

▲ What are chemicals themselves made of? (pages 8–9)

▼ What makes a metal a metal? (pages 26–27)

How can we all help to save chemicals, and reuse them as much as possible? (page 39)

▲ How can different chemicals be separated from a mixture? (pages 10–13)

▼ What chemical changes go on inside living things? (page 23)

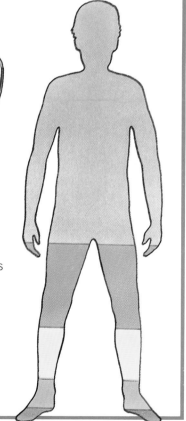

▶ Which chemicals are you made of? (pages 33–35)

Chemical World

"Chemicals" are not just the strangely colored bubbling liquids in the test-tubes of the scientist's laboratory. Chemicals are all around us, forming everything we smell, taste, touch, and see—and even things we cannot see, such as air or glass. Chemistry is happening all the time. Chemicals are changing all the time.

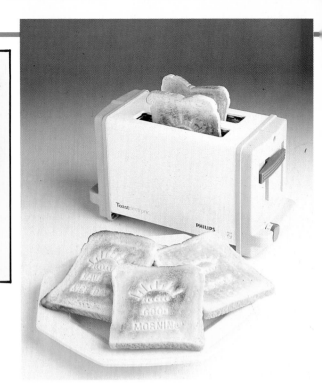

 ## Cooking Chemicals

The kitchen is one of the best places to see chemistry in action. We eat certain foods raw, but others must be cooked in some way. Cooking makes them safe to eat, or improves their taste, or makes them easily digested by the body. Cooking changes the chemicals in foods, usually by heat. As a piece of bread is heated in a toaster, its surface burns and becomes crispy, and its color changes to brown. A raw egg has a runny, slippery "white" and a thick, yellow liquid yolk. Boil it in water and both these substances turn solid, to produce a hard-boiled egg. Or heat a raw

▲ Toast is simply bread which has been "lightly burned" on each side. Burning usually changes

egg in a pan and stir it with butter and milk, and it changes to scrambled egg, which is light yellow and lumpy.

Browned toast made from white bread.

Hard-boiled egg

Scrambled egg

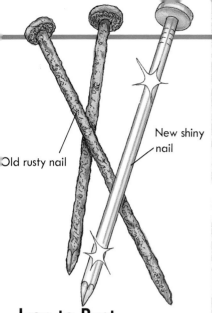

Old rusty nail

New shiny nail

Iron to Rust

▶▲ Iron, a metal (page 26), goes rusty if it becomes damp. The iron combines with water to make a new substance, reddy-brown rust. Leave a new iron nail in a dry place, such as a warm kitchen. Put other nails in a damp place, like a corner of the garden. Which ones go rusty? Old cars, beds, and many other objects have iron in them, and rust too.

Going Hot While Setting

Fresh plaster of Paris is soft and powdery. Mix it with water, leave it, and it becomes very hard. But the plaster does not "dry," in the way a towel dries on the clothes line. It goes through a chemical change when mixed with water, "sets" to produce a new substance, and produces heat. Mix some plaster with water in a plastic container. After 30 minutes, the side of the container should feel warm. **Caution:** Do not throw leftover plaster into the sink. It sets even under water!

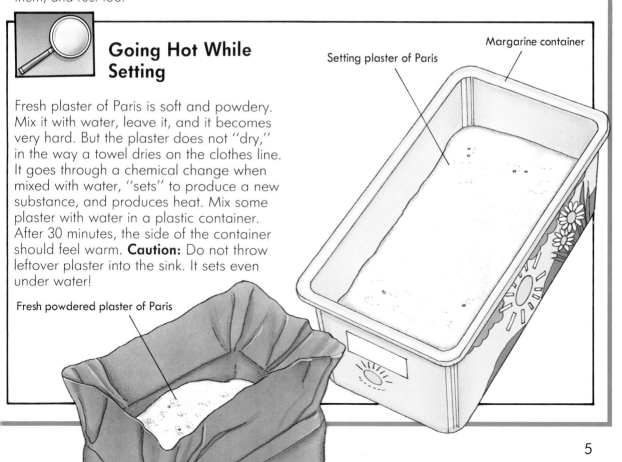

Margarine container

Setting plaster of Paris

Fresh powdered plaster of Paris

Chemical Properties

How can we identify a chemical? We can use our senses to detect its properties. We look at color and shininess. Provided a chemical is safe we can smell its odor, feel its hardness and texture and weight, and taste its flavor.

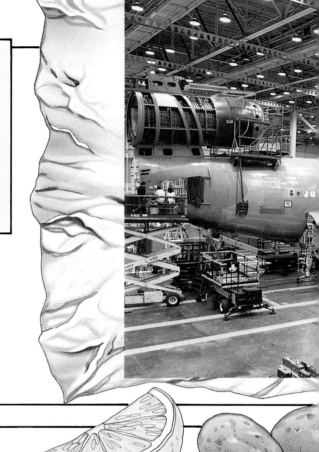

The basic properties of any chemical are always the same, no matter what the shapes of the objects made from it. Aluminum is a metal used to make "tin foil." It is also used, as much thicker sheets, to make the bodies and wings of aircraft.

▶ Floppy cooking foil seems to be made of a very different material from an aircraft. Yet they are both aluminum.

Testing Properties

Lemon   Potato

Chemicals may look and feel the same, but taste quite different. Ask a friend to grind up small amounts of sugar, salt, and chalk. Place them in separate piles on a plate. Which is which? Taste is the best way to tell them apart. **Caution:** Never taste an unknown substance. It could be poisonous.

There are two small pools of yellowish liquid chemicals on a plate. One is lemon juice, the other is cooking oil. How can you tell them apart? The easiest way is by touch. Rub them between your fingers. The cooking oil feels thick and slippery, and the lemon juice is thin and watery.

Taste

Feel

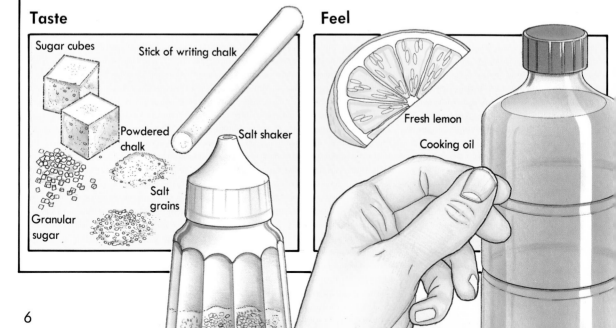

Sugar cubes

Stick of writing chalk

Powdered chalk

Salt shaker

Salt grains

Granular sugar

Fresh lemon

Cooking oil

Color

You could not color a picture with your eyes closed. Even if you knew where to draw you would not be able to choose the correct color of pencil without looking. Different colored pencils feel and smell the same. They can only be told apart by sight. You might end up with a very strange-looking picture indeed! Color is an important property of chemicals, because eyesight is an important sense for us.

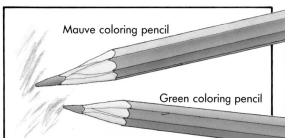

Mauve coloring pencil

Green coloring pencil

Apple

Salt

Sugar   Polystyrene

Chalk

A smell consists of tiny particles of odor chemicals floating in the air. As you sniff, the chemicals are detected by special nerves inside your nose. Substances which look and feel similar can sometimes be distinguished by smell. Cut small lumps of apple and potato. Give each a sniff to tell them apart.

Some chemicals are very heavy for their size, like the metal lead. Others are extremely light. Paint a pebble-shaped lump of expanded polystyrene to look like a stone, and place it next to a real stone. Ask a friend to lift each one. The friend will probably be surprised at the lightness of one of the "pebbles"!

Smell

Freshly-sliced apple

Freshly-sliced potato

Weight

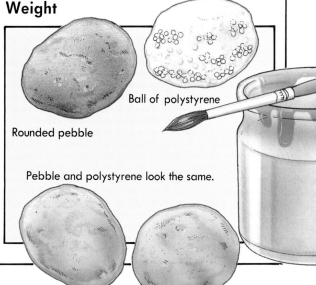

Rounded pebble

Ball of polystyrene

Pebble and polystyrene look the same.

Building Blocks

There are about 100 basic chemicals, called elements. Each element is made of building blocks called **atoms**. These are too small to see even under a microscope. All the atoms of an **element** are the same as each other, and they are different from the atoms of any other element. Atoms can be split into smaller particles, but these do not have the properties of the element.

Atoms, Molecules, and Elements

An element is a pure, single chemical. Iron is an example of an element. No matter how many times a piece of iron is divided, it still has the properties of iron—until, if you could carry on dividing for long enough, you are left with just one atom of iron. An atom is the smallest particle of the element that still has that element's properties. Atoms are rarely found on their own. Usually they are joined to other atoms, to form **molecules**. A molecule may be made of two atoms, or three, or many thousands. The atoms in it may all be of the same element, or from a few elements, or from dozens of different elements.

▼ One of the rarest elements on Earth is plutonium 238. It is used as a source of energy in space probes.

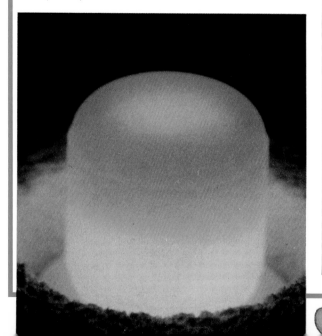

Making Atoms and Molecules

You can make some simple models of atoms, and combine them to give different types of molecules. Two of the commonest elements on Earth are carbon and oxygen. In chemistry, each element has a chemical symbol. The symbol for **carbon** is **C**, and for **oxygen** it is **O**. Just as atoms combine to form molecules, so their symbols can be joined to make a symbol for a molecule. For example, oxygen is in the air we breathe. But it does not occur as single atoms of oxygen, floating around. It exists as molecules of oxygen. Each molecule is made from two oxygen atoms joined together. The simple two-atom molecule of oxygen is written O_2.

Making Playing-Dough Atoms

"Atoms" of different colored balls of playing-dough can be linked by small sticks, to make different molecules. First, mix two cups of flour and one cup of salt in a pot. Add a few drops of food coloring to one cup of water, and stir this thoroughly into the mixture. Also stir in four teaspoons of cream of tartar and two tablespoons of cooking oil. Ask a grown-up to help you warm the pot gently over a low heat, stirring all the time. After the dough thickens, leave it to cool. Make two batches of dough: red for oxygen atoms, and black for carbon atoms.

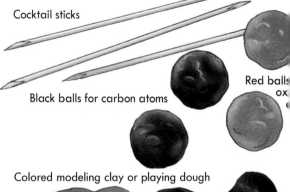

Cocktail sticks

Black balls for carbon atoms

Red balls ox

Colored modeling clay or playing dough

Equipment: Flour, salt, saucepan, food colouring, cream of tartar, cooking oil.

Carbon

There are several forms of the element carbon, depending on how the carbon atoms are joined together. One familiar form is soot. Wood, coal, heating oil, and similar substances contain lots of carbon (*page 30*).

Forms of carbon are graphite (the "lead" in a pencil) and diamond.

Carbon C

Oxygen

Pure oxygen is made by plants, during the process of photosynthesis. The plant's leaves capture the Sun's light energy and use it to make the plant's food. As this happens, molecules of oxygen are given off into the air.

Most living things need to breathe oxygen in order to stay alive.

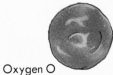

Oxygen O

Carbon Dioxide

Carbon dioxide is a waste product of our body chemistry. As we breathe out, it is released into the air. The harder our muscles work, the more carbon dioxide they make, and so we need to breathe harder and faster to get rid of it.

Carbon dioxide (CO_2) has one carbon atom and two oxygen atoms.

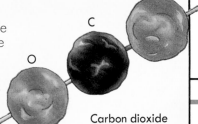

Carbon dioxide

Carbon Monoxide

A car engine combines carbon and oxygen to make carbon monoxide, a poisonous substance in exhaust fumes. Atoms can join only in certain ways, depending on how many links, or "bonds," they attach to each other.

Carbon monoxide (CO) has one carbon atom and one oxygen atom.

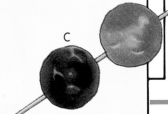

Carbon monoxide

Mixing, Dissolving

Some chemicals can mix together without change. Stir together powdered chalk and powdered iron (iron filings), and you would not notice any alteration. But other chemicals may react to produce a new kind of chemical (*page 14*). Or one chemical may dissolve in the other.

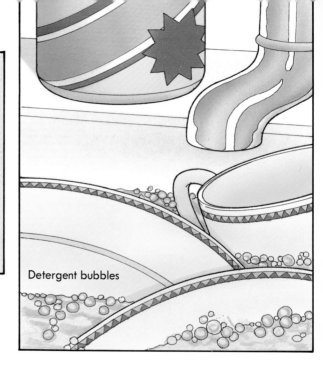

Detergent bubbles

Washing With Soap

In dissolving, a solid chemical "disappears" into a liquid one. We see this every day, when someone puts sugar in tea or coffee, or puts salt in water before cooking. The solid chemical (sugar or salt) is called the solute. The liquid one (tea or water) is the solvent. The two together are known as the solution. The solute alters its nature as it dissolves and "disappears" in the solution. This is because it changes from large groups of solute molecules into very small groups or even single molecules. The solute does not combine chemically with the solvent (*page 12*).

▼ Whether it is a dirty dish or a dirty car, soapy water helps to break up and dissolve the dirt and grime, and wash it away.

Which Chemicals Will Dissolve?

Water is the "Universal Solvent." Many substances dissolve in it, to make drinks, inks, cleaning fluids, and other useful solutions. You can test whether a substance dissolves by adding a small amount of it to a beaker of clean water, and stirring slowly. If the substance is soluble it will disappear and leave the water clear, though it may give its color to the water. Tea, coffee, salt, and honey should dissolve. Chemicals like sand and cooking oil do not dissolve.

Honey

Warm tap water

Coffee

Cooking oil

Sand

Salt grains

Tea bag

Which Dissolves Best—Hot or Cold Water?

This simple experiment shows that warm water is a better solvent than cold water. You need two identical beakers. Into one, put cool water straight from the cold faucet. Put very warm water from the hot tap into the other beaker. Then, add a teaspoon of salt to each beaker, and stir. The salt should dissolve and disappear into the warm water first. What happens if you keep adding salt to each beaker?

Teaspoon of salt grains

Very cold water

Very warm water

"Un-Dissolving"

When one chemical mixes or dissolves in another, the molecules of each chemical remain basically unchanged. It is possible to separate them again fairly easily. After separation, the chemicals are the same as they were at the beginning. When chemicals react together (*page 14*), separation is more complicated—and may be impossible.

Separating by Evaporation

A solute can be separated or "un-dissolved" from its solvent by evaporation. In this process, the solution is heated. The solvent gradually turns into a vapor or gas (*page 20*), leaving behind the solute. The vaporized solvent is changed from a gas back into a liquid by cooling so that it condenses. (**Condensation** is the opposite of evaporation.) Ask a grown-up to make a mug of steaming-hot coffee—a solution of coffee chemicals in water. Over the mug, hold a cold pot from the refrigerator. Water vapor condenses on the pot and turns to clear liquid water.

▼ Common salt is used for many purposes, from industrial processes to cooking. Sea water contains dissolved salt. This can be obtained by using the Sun's heat to evaporate sea water in shallow pools, called pans. The water slowly turns into a gas, to leave crystals of salt. The process is known as desalination.

Cold pot from refrigerator

Clear water collects

Lid of coffee jar

Mug of hot coffee

Separating Three Chemicals

How could you sort out a mixture of three chemicals: sand, salt and iron filings (powdered iron)? The answer is to use the chemical properties of each substance. Salt dissolves in water, but the other two do not. So add water to the

1 Dissolve the salt in water and evaporate this to obtain salt crystals.

2 Rub a magnet gently in the mixture to pick up the iron filings.

3 Only sand is left in the dish at the end of the experiment.

mixture, and the salt dissolves to form a solution. Pour this off carefully and leave it in a warm place. The water evaporates, to leave the salt on its own. Next, you need to remove the iron particles from the sand. Iron is magnetic, unlike sand. So gently push a magnet through the mixture and the iron filings stick to it. Tap them off into a another container, to leave only the sand.

Separating Colors

Many colored inks in felt-tip pens are mixtures of a basic set of colored dyes (*page 18*), in different amounts. Find out which dyes are in your pens by this simple experiment. (Use a pen with ordinary water-soluble ink, not the special alcohol-soluble kind.) Make a thick blob of ink on a piece of blotting paper. Prop this against some books so that the bottom of the blotting paper is in a jamjar of water, with the blob just above the surface. Water seeps up the paper and carries with it the different dyes, but at different speeds. The rainbow pattern reveals the colored dyes in your ink.

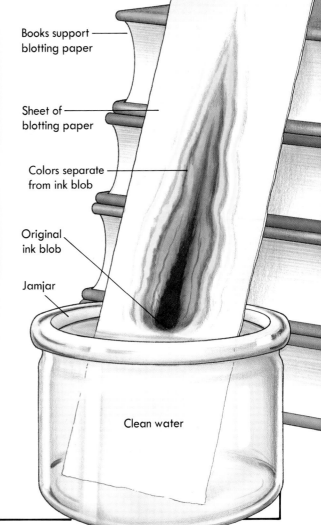

Books support blotting paper

Sheet of blotting paper

Colors separate from ink blob

Original ink blob

Jamjar

Clean water

Chemical Reactions

In a chemical reaction, substances that are mixed together react or combine in some way. They form a new substance, which is different from the original chemicals used to make it. Chemical reactions are going on all around us, as we drive a car, build a wall, cook a meal, or paint a door.

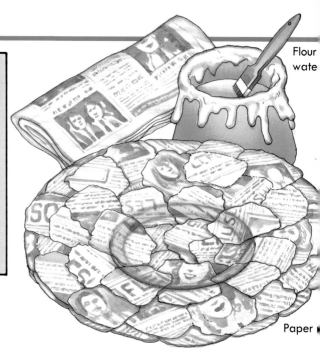

Flour
wate

Paper

Flour-and-Water Paste

Add two cups of flour to one cup of water, and stir to a stiff paste. They become a new substance which is sticky and gooey. This is a chemical reaction involving molecules of starch (*page 34*), one of the main chemicals in flour.

Playing-Dough Sculptures

A simple recipe for playing-dough is given on page 8. In fact, this is a recipe for a chemical reaction. The various ingredients, such as flour and salt, combine to make a new substance, the dough, which has quite different properties. The gentle heat needed to make the dough helps the chemicals to react together. In general, heat speeds up a chemical reaction, while cold slows it down. This is why we put foods in the refrigerator or freezer to preserve them, by preventing any chemical reactions. It is also why we heat foods in the oven, to make them undergo chemical reactions—what we call "cooking."

Playing-dough can be molded into many shapes, from model people to cups, cars, and volcanoes (*see opposite*). If you add some food coloring to the water before you make it, this will give an even color to the dough. Or you can put the finished models in a gently warm oven to dry them hard.

Flour-and-water paste sets hard. Paint layer of old newspaper with it and smooth them onto a mold, such as a plate. (Coat the plat with a thin smear of petroleum jelly, so the paper does not stick to it.) When hard, the plate can be painted in bright colors.

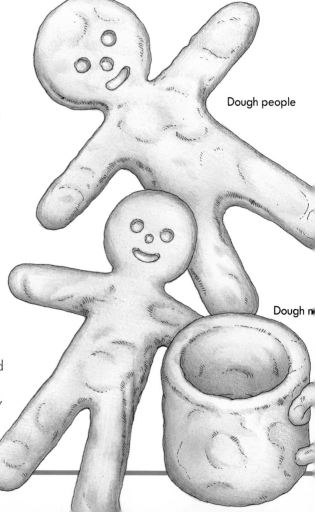

Dough people

Dough n

Fast and Slow Reactions

▲ A slow reaction, like the setting of cement (*top left*), takes several days. A fast reaction, such as gasoline vapor exploding in a car engine (*above*), happens in a fraction of a second.

It is often important to control the rate, or speed, of a chemical reaction. For example, setting cement undergoes a chemical reaction, like plaster of Paris (*page 5*). Imagine if cement set a few seconds after being mixed with water! Extra chemicals, "retardants," can be added to the cement so that it stays soft for longer. A fairly fast chemical reaction happens when baking powder is added to vinegar. You can make a model volcano using this reaction. Mold the sides of the volcano from playing-dough. Put a small pile of baking powder in the hole in the middle, with a few drops of some red food coloring. Then drip in vinegar. Foaming red "lava" pours from the volcano!

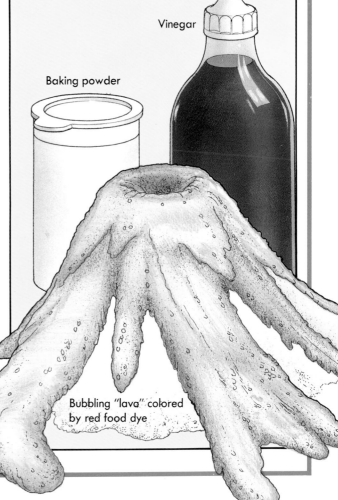

Vinegar

Baking powder

Drip vinegar into mouth.

Volcano mixture

Bubbling "lava" colored by red food dye

Acids and Bases

Chemicals have properties that we can detect, such as color and smell. Some also have another, more "chemical" property: they are either acids or bases. These chemicals are important. They react to produce a third group of chemicals, the salts. Acids, bases, and salts, and the reactions between them, are used in hundreds of household and industrial processes.

Testing for Acids

Acids are usually sour-tasting and corrosive, which means they "eat away" other substances. Even a weak acid, like vinegar or lemon juice, has a strongly sour taste. Powerful acids are so corrosive that they should never be touched, let alone tasted. If they splash on you, they will cause stinging pain and eat away your flesh! Acids change the color of certain chemicals, known as indicators. Bases (*see opposite*) turn the indicator a different color.

(1)

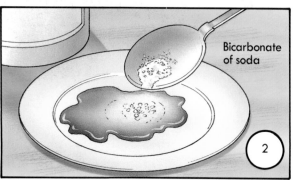

Bicarbonate of soda

(2)

(3)

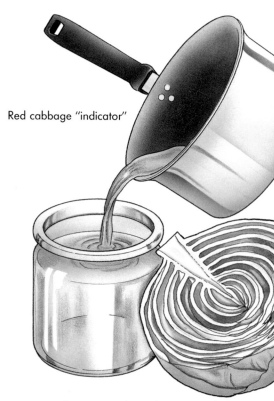

Red cabbage "indicator"

You can make a simple indicator at home, using red cabbage, and test it on some weak acids and bases. Ask a grown-up to chop up about half a red cabbage, bring it to the boil in water, stir well and leave it to soak for about 15 minutes. Pour the water through a sieve. This purple-blue water is your indicator (1). If you add a base to it, like bicarbonate of soda (*see opposite*), it should turn pale or greeny-blue (2). If you add an acid, like lemon juice, it should turn reddish or pink (3).

Make Mine Green and Fizzy!

A **base** is the "opposite" of an acid. It tastes bitter, rather than sour, and has a slimy feel. Powerful bases, such as drain-cleaning fluid, are as dangerous as acids and can corrode your skin. However, not all bases are this strong. A weak base is bicarbonate of soda, used in cooking, which can make ginger-bread rise. The chemical reaction between citric acid, in lemon juice, and bicarbonate of soda, a base, can put fizz into your drink.

Add a few drops of green food coloring to a pitcher of water. Stir in two tablespoons of confectioners' sugar and three teaspoons of bicarbonate of soda. Then add six teaspoons of fresh lemon juice. The acid lemon juice and bicarbonate base react to make carbon dioxide.

Home-made fizzy drink

Lemon juice

Food dye

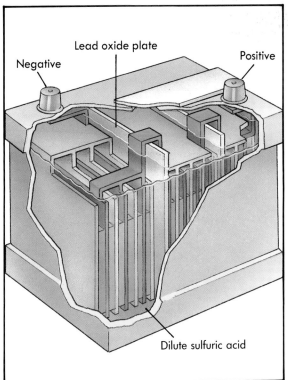

Negative

Lead oxide plate

Positive

Dilute sulfuric acid

Indigestion tablets

Tea leaves

Coffee beans

▲ A car battery contains sulfuric acid. Goggles should be worn when working with such batteries.

▲ Some medicines are weak bases. The substances obtained from plants, like tea and coffee, are bases called alkaloids.

Color Chemistry

Not all creatures can see in color. Humans can. Colors are very important to us, from the beauty of a fine work of art to the red warning light on a control panel. Color is a property of many chemicals. Some chemicals have especially strong or bright colors. We call these pigments dyes and colorings. They are used to give color to other objects, from a shirt to a ceiling!

▶ Materials such as wool and cotton can have their own natural colors. But we can change these by dyeing. Colored chemicals are used to stain the material.

Mixing Colors

This book would not look so bright and interesting if it was printed with just black ink, in "black-and-white." It would not give as much information, either, since it could not show the colors of things. Books printed in color use four inks, each containing strongly colored chemicals. The ink colors are magenta (pinky-red), cyan (blue), yellow, and black. Colored photographs and drawings are made up of tiny dots, each dot being one of these four colors. Your eyes "mix" these dots to give smooth areas of color. You can see the dots if you look at the pictures with a magnifying glass. Try mixing colored chemicals yourself, using your paint box.

▲ Paper being bleached before printing.

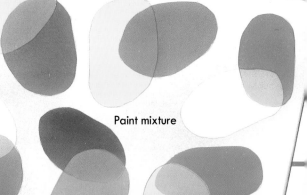

Paint mixture

Invisible Ink

A simple chemical reaction makes invisible ink visible. Squeeze some juice from a fresh lemon, then write, or paint it onto a piece of white paper. When the juice dries, your marks will be virtually invisible. To make them visible, ask a grown-up to put the paper in an oven at 350°F. for 10 minutes. The heat "burns" the chemical so it becomes visible.

Fresh lemon juice

Make a Space Meal!

We identify foods by their color, as well as their smell and taste. Peas are green, lemons are yellow, and oranges are—orange. You can trick your senses by using food colorings to alter ordinary foods, such as baked beans and eggs. Food colorings are chemicals which have a strong color and which have also been tested thoroughly to make sure they are safe to eat. Add a few drops of coloring to each food, to turn it a strange and unfamiliar color.

Green drink!

Red lemon!

Blue ice-cubes!

Purple baked beans!

Blue scrambled egg!

Changing States

In science, the **state** of a substance means whether it is solid, liquid, or gas. The same chemical can exist in each of these states, depending on its temperature and the pressure on it. A familiar example is water. When very cold, it is solid ice. Normally, it is liquid water. When hot, it is gaseous water vapor. Can you think of other common chemicals that change state?

Solid, Liquid, Gas

In a solid, the atoms or molecules of the substance are arranged in a certain pattern and cannot move. In a liquid, the pattern of arrangement has broken down. The molecules are free to move, but they still stay near each other because they attract each other, like a magnet attracts iron. In a gas, the molecules are much farther apart, too far to attract each other. The "crazy string" used at parties changes states as you squirt it out. Inside the can it is in liquid form, under high pressure. As it squirts out, the pressure is released, and the liquid comes into contact with the air. It changes into a bendy, stringy, plastic solid.

▼ We use the chemical carbon dioxide in all its states. As a gas, it forms the bubbles in fizzy drinks. As a liquid, it is stored in soda-siphon refills. (A liquid takes up much less space than a substance in gas form.) As a solid, carbon dioxide "snow" keeps things extremely cold.

Skating On Water, Not Ice

Pressure of skate blades on ice melts it into water.

One way to turn a solid to a liquid is by heating it. If you warm ice, it changes to liquid water. We call this "melting." Ice changes to water at 32°F. This is its melting point. Solid carbon dioxide has a melting point of minus 109°F. Another way to turn a solid to a liquid is to increase its pressure. Put ice under great pressure and it melts into water, because the pressure causes its temperature to rise. This happens under an ice-skater. The skates' narrow blades press on the ice as they move along and melt it into a tiny pool of water. The slippery water lets the blades slide past.

Non-Melting Chemicals

To stop chocolate melting too quickly, put a few pieces in the freezer or icebox for a couple of hours. Trick your friends by giving them warm pieces, which have been in the warm kitchen, while you have the cold pieces. The warm pieces soon melt all over their hands, while your pieces stay cool and do not melt. This is because a warm piece of chocolate needs only a small amount of heat, which it takes from the hands, to raise its temperature to melting point. A very cold piece needs much more heat to make it melt, and this takes longer to pass from your hands to the chocolate.

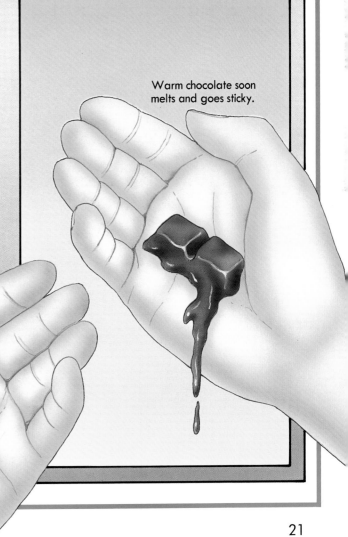

Warm chocolate soon melts and goes sticky.

Ice-cold chocolate stays hard and solid.

Make an Eggshell Fire Extinguisher

Fire involves chemical reactions that need oxygen, a gas in the air (*page 8*). If there is no oxygen, the fire goes out. Fire blankets and some fire extinguishers work by smothering a fire, to prevent oxygen reaching it.

One substance that can be used for smothering is a gas, carbon dioxide (*page 9*). You can make a small extinguisher using eggshell and vinegar. Drop small pieces of a clean eggshell into a few teaspoons of vinegar, in a deep container such as a plastic beaker. The ethanoic acid in vinegar reacts with a chemical called calcium carbonate in the shell, to make carbon dioxide gas. This slowly fills the beaker, though you cannot see it. Use it to smother a candle, as shown below.

▲ In an emergency, a carbon dioxide (CO_2) fire extinguisher soon smothers a fire and suffocates the flames.

Carefully pour gas over small lit candle.

Stand candle in sand in deep bowl.

Balloons filled with air containing carbon dioxide – from your lungs!

Collecting Oxygen

Plants and animals make carbon dioxide during their bodily chemical reactions, and plants also make oxygen, by photosynthesis (*page 9*). You can collect these gases as shown here. Use a water plant such as Canadian pondweed, in a bowl or trough of water. Put a large upturned jamjar over it, supported by thread spools. The plant must be in full sunlight for the best effect. Over a few hours, bubbles of gas form on its leaves and float up to collect in the jamjar. Are they carbon dioxide or oxygen? Carefully lift the jamjar, turn it over, and ask a grown-up to put a long, lighted taper into it. Carbon dioxide will smother the flame. Oxygen will make it burn more brightly.

Gas bubbles collect in jamjar.

Upturned jamjar

Canadian pondweed (or similar waterweed)

Glass or plastic trough

Clean water

Spools

Weed in water

Un-Freezeable Water

Pure water changes from a liquid to solid as its temperature falls below 32°F. We call this freezing. But water that contains a dissolved chemical, such as salt, freezes at a lower temperature. You can make "un-freezeable water" by dissolving as much salt as you can in cold water. Put this salt solution in one plastic container, and clean water in a similar container. Place them in the icebox or freezer. The pure water soon turns to solid ice. The salty water stays liquid for much longer. In very cold weather, salt is spread on roads. It dissolves in any water there, to stop dangerous ice forming on the road surface.

▶ The water in an iceberg freezes at 32°F since it is made from rain water. Sea water freezes well below 32°F, because of the salt dissolved in it.

Mixing Water and Oil—Eventually

Water and cooking oil are both liquids. But they are very different chemicals. It is difficult even to mix them thoroughly. This experiment shows that they can be mixed, eventually, with the help of another liquid chemical—liquid soap.

Water and oil do not mix because their molecules are different. A molecule of water has two atoms of hydrogen and one of oxygen, written H_2O. But a molecule of oil is made of dozens or hundreds of atoms, and contains carbon as well as hydrogen and oxygen. If you shake oil and water together, the oil splits into small globules, but it does not truly dissolve. This mixture, of tiny blobs of one liquid floating in another liquid, is called an **emulsion**. Detergents help to change the oil molecules so that the oil blobs split up even further. They become so small that the oil seems to "dissolve" in the water.

Cooking oil

Pour oil gently onto water.

1

2

Oil

Water

▲ Put about 2 inches of water in a narrow clear jar or bottle. Carefully add the same amount of oil (1). This soon settles out separately and floats on the water (2).

How Pepper Walks on Water

The surface of a liquid, where it touches the air, has special chemical properties. One of these is surface tension, which creates a kind of "skin" on the liquid. To show water's skin, put some water into a clean plate or shallow tray. Sprinkle pepper grains from a pepper-pot evenly onto the surface. The grains float and stay still on the surface. Gently touch the water with a bar of soap. At once, the pepper grains should float away from the soap and collect around the edges of the container.

How it works: Soap changes the surface tension. The grains are pulled away by the surface tension around the sides.

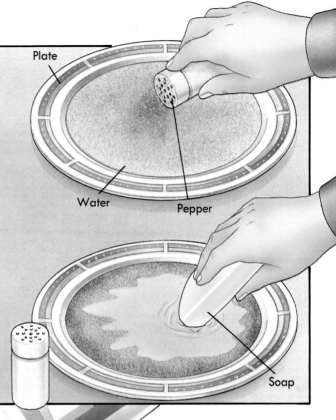

Plate

Water   Pepper

Soap

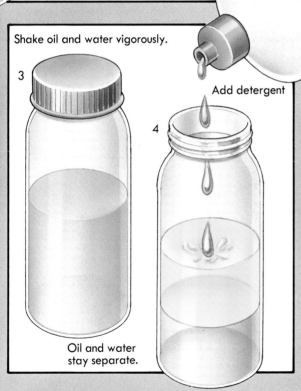

Shake oil and water vigorously.

3

Oil and water stay separate.

Add detergent

4

▲ Put a top on the jar and carefully shake it. The oil breaks into tiny blobs to form a milky mixture (3). But in an hour, it settles on the water again. Add a few drops of detergent (4).

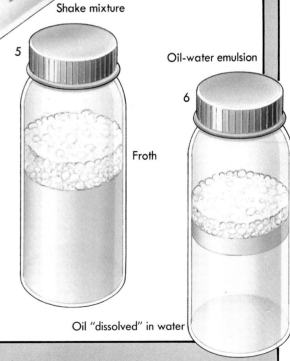

Shake mixture

5

Oil-water emulsion

6

Froth

Oil "dissolved" in water

▲ Shake the jar once more, and a milky liquid forms (5). This has a frothy top, a layer of yellowish milky liquid, and—eventually—a layer of oil "dissolved" in water (6).

Metals

Some of the most familiar chemicals are metals. Iron, aluminum, lead, and steel are all types of metals. The first three of these are also elements (*page 8*). Steel is a combination of iron with small amounts of other substances, such as carbon. Most metals are hard, shiny, strong, and carry electricity. They are an essential part of our modern world, used to make hundreds of machines.

The World of Metals

Iron, chemical symbol Fe, is one of the most common elements in the Earth's rocks. We see iron as a hard metal that is fashioned into kitchen utensils, railings, gates, and girders. Yet deep inside the Earth, where temperatures and pressures are enormous, there are vast amounts of liquid iron.

Aluminum, symbol Al, is even more common than iron. It makes up about one-twelfth of the Earth's surface. It is very light, used to make parts for planes and high-speed cars and trains.

Lead, symbol Pb, is one of the heaviest and softest metals. In former times lead was beaten into sheets for roofs and bent into pipes for water. However we now know that lead is poisonous to the human body, even in tiny amounts.

Titanium, symbol Ti, is another very common metal in the Earth's rocks. But it is difficult to obtain in its pure form. Strong as steel, and half the weight, it is used in jet engines.

Copper, symbol Cu, is a reddish metal that carries electricity very well. It is suited to making electrical wires and cables. Also it does not rust, and has taken over from lead in water pipes.

Alloys are combinations of metals. Bronze is an alloy of copper and tin. Brass is an alloy of copper and zinc.

Metals Around the Home

Look around your home or school. Can you spot things that are made of metals? Some are shown below. Each type of metal is chosen for hardness and strength, to suit its job, from a paper clip to a carving blade. Most metals are found in nature in rocks, combined with many other chemicals. They must first be extracted, which means obtaining them in a pure form. Then they are heated to make them melt, and poured into molds.

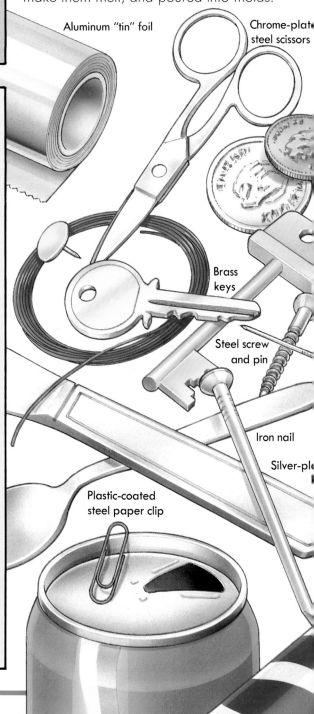

Aluminum "tin" foil

Chrome-plated steel scissors

Brass keys

Steel screw and pin

Iron nail

Silver-pla

Plastic-coated steel paper clip

Sorting Metals

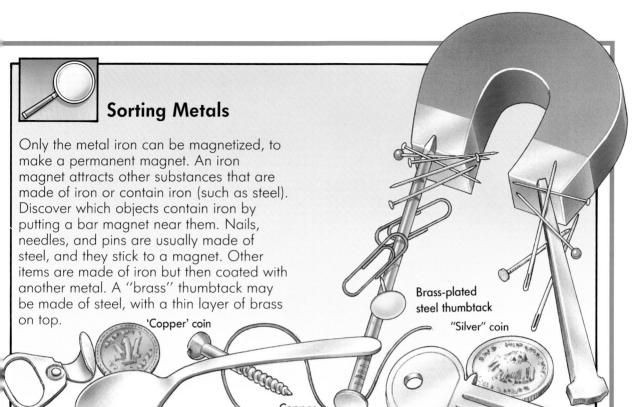

Only the metal iron can be magnetized, to make a permanent magnet. An iron magnet attracts other substances that are made of iron or contain iron (such as steel). Discover which objects contain iron by putting a bar magnet near them. Nails, needles, and pins are usually made of steel, and they stick to a magnet. Other items are made of iron but then coated with another metal. A "brass" thumbtack may be made of steel, with a thin layer of brass on top.

'Copper' coin

Brass-plated steel thumbtack

"Silver" coin

Copper electrical wire

Making Metals

Certain metals are heated or "blasted" in a furnace to extract them from their ores (the rocks that contain them). Iron flows straight from its furnace at a temperature of 2,730°F. In some cases the heat is created by passing huge electric currents through, rather like a gigantic toaster! The rocks melt and the metal separates and floats to the surface.

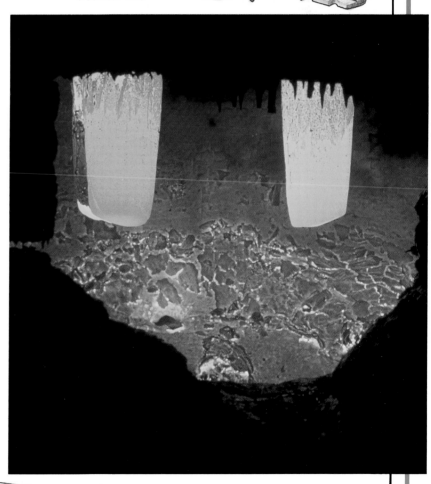

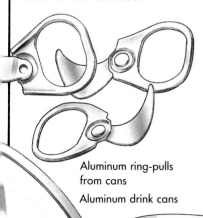

Aluminum ring-pulls from cans

Aluminum drink cans

27

Chemical Electricity

Electricity is generated when certain chemicals react together. We use chemically made electricity to power many machines, from a flashlight to a personal stereo, an electronic quartz watch, and the electrical circuits of a car. We usually call these devices "batteries." When the chemicals are used up, the electricity stops—unless the battery is rechargeable!

▼ A small battery as used in cameras and watches.

Make a Lemon Battery

A simple battery which produces only a small, safe amount of electricity can be made from a lemon, as shown below. In a battery, there is a central chemical called the **electrolyte**. This is placed between, and in contact with, two other substances, the **electrodes**, which are usually made of metal. When a wire and bulb are added to make an electrical circuit, the chemical reaction between the electrolyte and the electrodes creates electricity. This flows around the circuit and makes the bulb glow. In an ordinary flashlight battery, or "dry cell," the electrolyte is ammonium chloride paste. One electrode is a carbon rod in the center of the battery, with a brass cap. The other is the zinc battery case.

To make the lemon battery, stick a brass thumbtack and a steel paper clip into opposite sides of a fresh lemon. Connect a low-voltage torch bulb (less than 3 volts) by two pieces of electrical wire to make the circuit. The acidic juice of the lemon is the electrolyte. The thumbtack and the paper clip are the two electrodes.

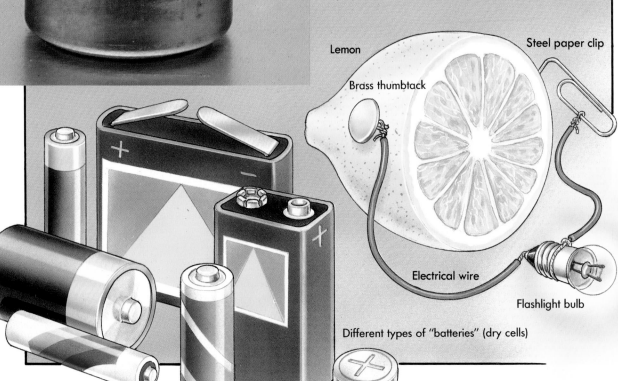

Lemon

Steel paper clip

Brass thumbtack

Electrical wire

Flashlight bulb

Different types of "batteries" (dry cells)

Electricity in Industry

Electrical power is vital in today's world, from powering toy trains to real ones. The metal aluminum is obtained in pure form by passing huge amounts of electricity through its refined rock ore in pots, in a process known as **electrolysis** or "smelting" (*see below*). Each smelting pot uses electrical currents of up to 150,000 amps, and there may be 1,000 pots in one refinery. The electricity for this process may well have come originally from another chemical reaction—the burning of coal or oil.

Splitting Up Water

Chemical changes can generate electricity. Electricity can also produce chemical changes. Water is a simple chemical made of hydrogen and oxygen (*page 24*). If an electric current is passed through it, between electrodes, the water is split into its two constituents, hydrogen and oxygen, which are both gases (*see below*). This process, electrolysis, is used in industry in many ways, such as obtaining metals like aluminum (*above*). If one of the electrodes is a metal, it will become covered or "plated" with any metal in the electrolyte. This is how cans are made.

Connect a battery using wires to two HB pencils, which are the electrodes. Place their other ends in water with a little salt dissolved in it, as shown. The electricity splits water into hydrogen and oxygen, which collect as bubbles around each pencil tip.

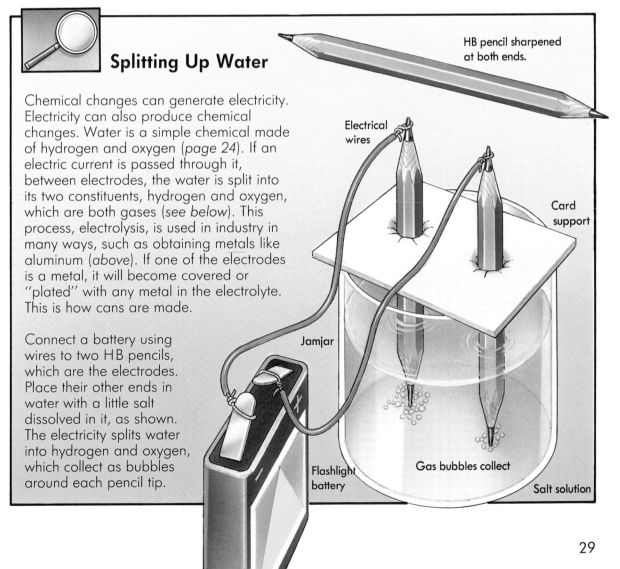

HB pencil sharpened at both ends.

Electrical wires

Card support

Jamjar

Flashlight battery

Gas bubbles collect

Salt solution

Carbon Chemistry

Carbon-containing chemicals are the basis of life. Atoms of carbon combine with atoms of other elements, like oxygen, hydrogen, and nitrogen, to make the bodies of living things. Every plant and animal, from a daisy to an elephant, has a body whose chemicals are based on carbon. The study of carbon chemicals and how they react is known as **organic chemistry**.

The Power for Life

Living things need energy to power the chemical reactions going on inside their bodies. This energy comes from the Sun. Plants capture the Sun's light energy directly, by photosynthesis (page 9). They use some of this energy for living. They also store some in chemical form, as the molecules in their bodies. Animals eat plants and take in the energy-containing molecules. They use some of the energy for their own life processes, and store the rest in their bodies. Other animals eat them, take in the energy, and so on. In this way we can trace the web of life back, to see that the energy for all life comes from the Sun. The Sun's heat warms the Earth and causes winds—another form of energy. In the Sun, the vast amounts of energy are created by a type of massive chemical-nuclear reaction. The Sun will continue to shine for billions of years, before it runs out of energy.

Over millions of years, the Sun's energy has been stored in countless living bodies. By slow chemical reactions, they turned into coal, oil, and gas. When we burn these fuels, we release this energy.

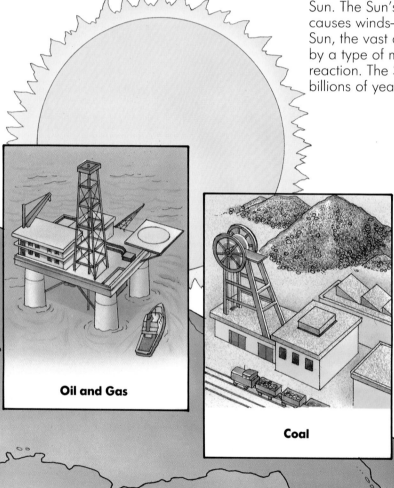

Oil and Gas

Coal

Wind and Sun

Toasting Marshmallows

Most of the foods we eat were once living things, either plants or animals. So they contain lots of carbon. Often, when organic chemicals are burned, the carbon changes its form and may be left in its powdery black form. This is why food turns dark or black if it's burned during cooking.

Ask a grown-up to toast marshmallows safely near a gentle flame. You can see the sugary marshmallows burn slightly, and carbon appears as a dark substance.

Before toasting   After toasting

Turn Milk into Plastic

Many plastics are made from petroleum oil. Oil formed in the rocks over millions of years, from the bodies of billions of small sea creatures. You can make a similar "plastic" in a few minutes—using milk, another organic (carbon-containing) substance.
Ask a grown-up to warm some creamy milk in a pot. When it is just simmering, slowly stir in a few teaspoons of vinegar. The acidic chemicals in the vinegar react with the organic milk chemicals. Keep stirring until it becomes rubbery. Let it cool and wash it under running water. You now have your own plastic, which you can bounce around or mold into shapes.

Drip vinegar slowly into warm milk.

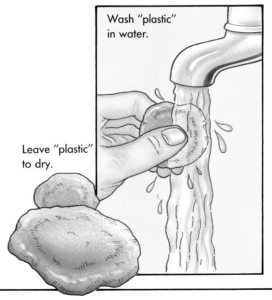

Wash "plastic" in water.

Leave "plastic" to dry.

Biochemistry

Biochemistry is the study of chemical reactions in living things. Experts have learned much about how biochemical reactions occur and how they are controlled. Living things such as microscopic bacteria can be programmed to act as tiny "biochemical factories" and make useful substances, such as medicines (*page 37*).

▶ In this fermenting room, useful chemicals are being made biochemically, by microscopic bacteria and yeasts. Cleanliness is vital. Unwanted germs can ruin the reactions.

Biochemicals Eat a Boiled Egg!

"Biological" washing powders contain enzymes. Inside a living body, there are thousands of enzymes. They control the speed of biochemical reactions going on in the body. Some enzymes are especially good at splitting up or "eating" organic substances. In biological washing powder, these types of enzymes attack and dissolve away oil, and dirt stains. Dissolve ordinary washing powder in warm water in a jar, and biological powder in another. Put a peeled hard-boiled egg into each jar, and leave them in a warm place for a few days. Enzymes in the biological powder "eat away" the organic chemicals of the egg, while the other egg is not affected.

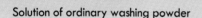

Solution of ordinary washing powder   Solution of "biological" washing powder

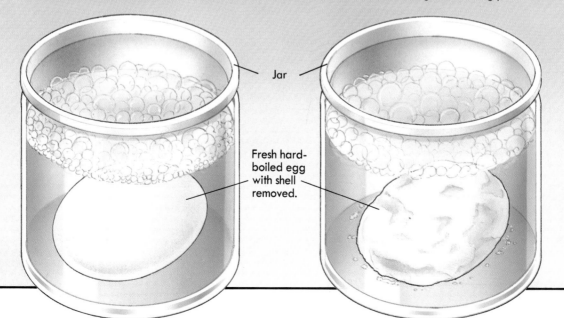

Jar

Fresh hard-boiled egg with shell removed.

The Conditions for Life

Living things need certain conditions in order to grow and flourish. They need a supply of oxygen, the right temperature, and the right amounts of moisture. These conditions allow the biochemical reactions of life to continue inside their bodies. Each plant and animal is suited to particular conditions. A waterweed needs to be in water and a mouse must be in air, or they will die.

Dry cotton balls

Moist cotton balls

Soaking-wet cotton balls

Even small changes in the chemical conditions surrounding a plant or animal, can cause problems. This simple experiment grows cress seeds under different conditions. You need three plastic jar lids. Put a layer of cotton balls into each, and sprinkle on a few cress seeds. Place them in a warm, light, airy place. Keep the cotton balls dry in one lid, damp in the second, and wet in the third. Which grows best?

Chemicals in the Body

Like everything in the world, the human body is made up of chemicals. And like all living things, you are made mainly of organic chemicals such as proteins, fats, and carbohydrates (page 35). These are based mainly on the elements carbon, hydrogen, and oxygen. However, the chemical reactions for life cannot happen unless these organic molecules are free to move about, mix, and react together. To do this, they must be dissolved or floating in water. This is why there is so much water in the body. In fact, you are almost two-thirds water!

Water 65%

Protein 18%

Fat 10%

Carbohydrate 5%

Others 2%

Recipe for the Body

The diagram above shows the proportions of some of the main chemicals in the body. There are also dozens of other chemicals, present in smaller amounts. They include iron for your blood, sodium and potassium for your nerves, calcium for your bones and teeth, and phosphorus for turning energy into movement.

The Chemicals in Food

When we mention "chemicals" in food, people may imagine the artificial colorings and flavorings made in a chemical factory. But food is made up of many other natural chemicals, that are part of the plants and animals it is made from. To stay healthy, your body needs the right proportions of three main types of food chemicals. Carbohydrates are found in sugary and starchy foods, such as bread, fruits, and vegetables. They are the main energy-providers for the body. Fats and oils occur in dairy products, fatty meats, and oily foods. They are used to make certain parts of cells. Proteins are contained in meat, fish, dairy products, fruits, and vegetables. They provide most of the building materials for the body, for growth, and repairing wear-and-tear.

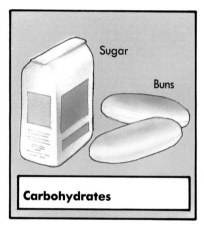

Sugar

Buns

Carbohydrates

Oil

Butter

Margarine

Fats and Oils

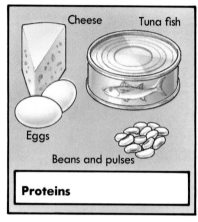

Cheese

Tuna fish

Eggs

Beans and pulses

Proteins

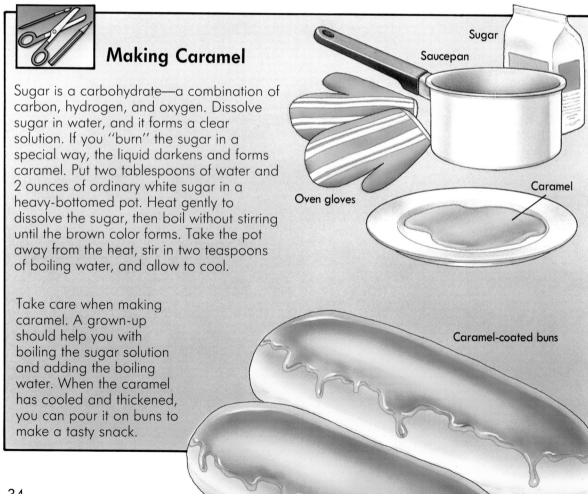

Making Caramel

Sugar is a carbohydrate—a combination of carbon, hydrogen, and oxygen. Dissolve sugar in water, and it forms a clear solution. If you "burn" the sugar in a special way, the liquid darkens and forms caramel. Put two tablespoons of water and 2 ounces of ordinary white sugar in a heavy-bottomed pot. Heat gently to dissolve the sugar, then boil without stirring until the brown color forms. Take the pot away from the heat, stir in two teaspoons of boiling water, and allow to cool.

Take care when making caramel. A grown-up should help you with boiling the sugar solution and adding the boiling water. When the caramel has cooled and thickened, you can pour it on buns to make a tasty snack.

Sugar

Saucepan

Caramel

Oven gloves

Caramel-coated buns

34

Soda Bread

Normal bread uses the biochemical reactions in yeast, to produce the carbon dioxide gas which makes it "rise" (see below). In soda bread, a chemical reaction between baking soda, sugar, buttermilk, and other ingredients produces carbon dioxide bubbles in the soft dough. As the dough is heated further, it sets like a sponge.

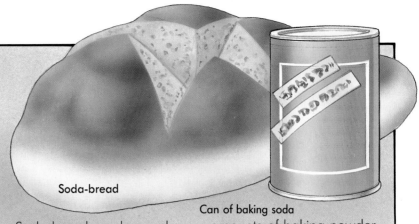

Soda-bread

Can of baking soda

Soda bread can be made at home, with the help of a grown-up. Find the recipe in a good cookery book, and follow it carefully. It is important to add the exact amounts of baking powder and baking of soda. Too little, and the bread will not rise when baked. Too much, and it will bubble up quickly and then collapse.

Wine and Bread

Yeast consists of tiny living cells, related to mushrooms. As they "breathe" they make carbon dioxide gas. To make bread, yeast cells are mixed into the dough. The cells create bubbles of carbon dioxide gas, which puff up the dough and make it rise. Yeast cells also make alcohol as a by-product.

Block of dried, pressed yeast

Cheeses and Yogurts

Cheese is made by a two-stage chemical change. The first involves milk and rennet. This is an enzyme found in calves' stomachs. Rennet digests or "curdles" the milk and makes it partly solid. The curdled milk is warmed, squeezed, shaped into blocks and stored. During this time bacteria turn it into ripe cheese or yogurt.

Cheese and yogurt

The Chemical Industry

In olden times, people's needs were simple. They used natural materials and ate fresh foods, or foods stored by drying or salting. Today we have invented any number of unnatural, man-made chemicals that are used to manufacture cars, planes, stereo systems, toys, buildings, furniture, and machinery. Our food has chemicals added. The chemical industry's products are a major part of everyday life.

► ▼ A giant refinery (*right*) makes dozens of chemicals out of crude oil, from gasoline for cars to asphalt on roads. A robot (*below*) paints a new car.

Robot arm

Agrichemicals

Farmers rely more and more on artificial chemicals to grow their crops. Man-made fertilizers are added to the soil, to improve yield. Seeds are coated with fungicides when they are stored, to stop them going bad. The soil is sprayed with herbicides, which kill the unwanted plants we call "weeds." Growing crops are sprayed with pesticides, to stop insects and other pests from damaging them.

▶ A sprayer soaks the growing crops with a cocktail of chemicals, designed to kill pests and stop diseases.

Biochemicals

The biochemical industry has made enormous progress in recent years. Bacteria and other microscopic organisms (*page 32*) can be altered to manufacture the chemicals we need. In former times, many medicines were extracted from animals and plants. These medicines could be difficult to obtain, and expensive. Experts in biotechnology can now produce some of these medicines in "fermenting vats." Insulin, a hormonal medicine needed by people with the disease diabetes, was purified from the leftover carcases of cows and pigs. Human-type insulin can now be made in a fermenting vat.

◀ Biotechnologists and biochemists work to develop new ways of manufacturing medicines, drugs, and food ingredients.

Waste Chemicals

The chemical business is big business. Huge amounts of money and many thousands of jobs are involved. But chemicals, especially waste chemicals, are damaging our world. They are soaking into the water, seeping through the land, and spreading into the air. How will they change the world? For example, we are discovering that chlorofluorocarbons (CFCs), chemicals used in some aerosols, are damaging the ozone layer around the Earth. Dozens of other chemicals might cause similar problems in the future. No one can be sure how these substances will affect our land, sea, and air.

Damaging waste chemicals take many forms. Some are radioactive. Used-up fuels, protective clothing, and equipment from nuclear power stations give off radiation that could damage life and cause disease. The radiation may last for thousands of years in some cases. No one knows how to deal with these chemicals and make them safe. At present, we are storing them as best we can.

▲ More and more people are becoming aware of the dangers from waste chemicals. This is especially so when the chemicals are not disposed of properly.

 SOS: Things To Do

We can start a Save Our Surroundings by being careful with chemicals of all kinds.
● Buy only what you need. A seemingly harmless product, like a little plastic toy, needs energy and raw materials to make it.
● Check before buying. Chemicals which do not damage the surroundings too much are usually labeled "environment-friendly" or similar. This is particularly important for strong chemicals such as detergents and cleaners.
● Use chemicals carefully. Do not buy more than you need, letting the rest go bad. Store chemicals safely. Dispose of them according to the instructions. If you do not need all you have bought, sell or give away the rest.
● Recycle chemicals. Collect glass, paper, aluminum, and other substances so that they can be re-used (see opposite). Buy things made from recycled chemicals if you can.
● Help to save the environment. Join an organization that does practical work such as recycling or cleaning up natural areas.

◄ ▼ Keep the countryside tidy. Waste and litter are not only unsightly, they are also a danger to wildlife and a waste of natural materials. We cannot go on wasting chemicals for ever. Supplies will run out.

Aluminum cans

Papers

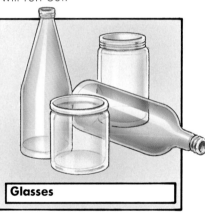

Glasses

Recycling

It is extremely wasteful to extract huge quantities of raw materials from the Earth, only to throw them away, little changed, after use. Many substances can be re-used and recycled, to save both materials and money. Have your own "recycling bins" at home, and visit the bottle bank or paper-collection point often.

Galvanized steel cans

Plastics

Index

Page numbers in *italics* refer to illustrations, or where illustrations and text occur on the same page.

Editor: Thomas Keegan
Designer: Ben White
Illustrators: Kuo Kang Chen
                Peter Bull
Consultant: Terry Cash

Cover Design: Pinpoin Design Company
Picture Research: Elaine Willis

Photographic Acknowledgments
The publishers would like to thank the following for kindly supplying photographs for this book: Page 6 J.F.P. Galvin: 7 ZEFA; 9 Allsport/Simon Bruty; 12 ZEFA (top) Science Photo Library (bottom); 13 Frank Lane Picture Agency; 14 Science Photo Library; 18 ZEFA; 24 ZEFA; 25 ZEFA; 27 Science Photo Library; 29 Crown Copyright/Reproduced with the permission of the Controller of Her Majesty's Stationery Office; 32 The Hutchison Library; 38 South American Pictures; 39 ZEFA (top) Science Photo Library (bottom).

Published in 1990 by Warwick Press,
387 Park Avenue South, New York, N.Y. 10016.
First published in 1990 by Kingfisher Books.
Copyright © Grisewood and Dempsey Ltd. 1990

All rights reserved.
Printed in Hong Kong

Library of Congress Cataloging-in-Publication Data

Parker, Steve.
    Chemistry/Steve R. Parker.
        p.   cm.—(Fun with science)
    Summary: Illustrates basic principles of chemistry, using experiments and tricks and showing things that can be made.
        ISBN 0–531–19085–4
        1. Chemistry—Experiments—Juvenile literature.
    [1. Chemistry—Experiments.   2. Experiments.]
    I. Title.   II. Series.
    QD38.P33   1990
    540—dc20   90–12028
                                                        CIP
                                                        AC